D1281632

STEAM & Me™
DINOSAURS

L. J. TRACOSAS

Starry Forest Books

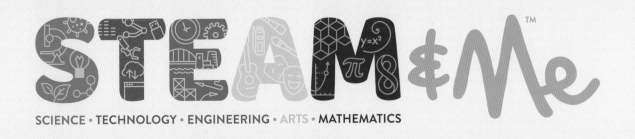

SCIENCE • TECHNOLOGY • ENGINEERING • ARTS • MATHEMATICS

Draw a super-smart robot. Create your own wind energy. Find out if your teeth are as sharp as a shark's. Go back in time to the world of dinosaurs or rocket into space. Power up that scientific brain of yours with **STEAM&Me**!

Photos, facts, and fun hands-on activities fill every book. Explore and expand your world with science, technology, engineering, arts, and math.

STEAM&Me is all about you!

New ideas sure to change how you see your world

Fascinating facts to fill and thrill your brain

Dinosaurs on the move!

Some dinosaurs walked on two legs. *T. rex* had little arms but could run on its two very big and strong legs. Some dinosaurs, like *Triceratops*, walked on four legs and moved like a rhinoceros. Other dinosaurs could move on either two legs or four legs.

Iguanodon walked on four legs but could also stand up on two legs.

Zoom!

Little *Compsognathus* might have been able to run at speeds up to 40 miles per hour, which is as fast as a motor scooter.

How fast can you go? Get down on your hands and knees. Crawl on the ground as fast as you can to the other side of the room. Now get up and walk the same distance. Are you faster on four legs or two?

STEAM&Me

Hands-on activities to spark your imagination

Great photos to help you get the picture

Ready to go back in time?

Imagine traveling back in time millions of years to when dinosaurs roamed Earth. You'd see some huge dinosaurs: some of the biggest creatures that lived on Earth. You'd see small dinosaurs no bigger than turkeys. Some dinosaurs were gentle plant-eaters. Others were fierce meat-eaters. All of them were amazing—and then all of them disappeared. Let's learn about the dinosaurs and the world they lived in.

Old Dinos, New Ideas

Luckily for us, dinosaurs left traces called **fossils**, which are parts or prints of plants or animals, found in earth or rock. Scientists are still finding new fossils, so we're still learning new things about dinosaurs.

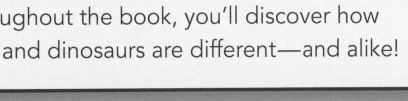

Dynamic dinos! In boxes like these throughout the book, you'll discover how you and dinosaurs are different—and alike!

STEAM & Me

Diplodocus
lived 161 million
to 146 million
years ago.

Stegosaurus had plates running the length of its backbone. All reptiles have backbones. Humans like you do, too.

Are you cold-blooded or warm-blooded? Touch your skin. Does it feel warm? When you go outside, your body temperature stays mostly the same, whether it's a hot day or there's a blizzard. That's because humans are warm-blooded.

STEAM & Me

What were dinosaurs?

Dinosaurs were reptiles that existed a long time ago. Like all reptiles, they had a backbone, they breathed air, and they were cold-blooded. Lots of reptiles exist today. Alligators, lizards, snakes, turtles, and crocodiles are all reptiles.

Brrr...

Reptiles are animals that are cold-blooded. That means their body temperature changes depending on how hot or cold it is outside. You may have seen a lizard, turtle, or snake soaking up the sun. Reptiles do that to get warm because their bodies can't warm themselves.

Dinosaurs lived during the Mesozoic era.

The Mesozoic **era** was divided into three time periods: Triassic, Jurassic, and Cretaceous. The age of dinosaurs began more than 245 million years ago, during the Triassic period. The Earth the dinosaurs inhabited was different from Earth as you know it now.

During the time of the dinosaurs, Earth was warmer. Larger plants grew, and the oceans were bigger.

Triassic

Coelophysis was a Triassic-period dino. Dinosaurs first appeared during the Triassic period.

Jurassic

Some of the biggest dinosaurs like *Apatosaurus* lived during the Jurassic period.

Cretaceous

Triceratops and *Tyrannosaurus rex* were Cretaceous-period dinosaurs. Dinosaurs disappeared from Earth at the end of the Cretaceous period. That was about 65 million years ago.

Say Hello to Sue

Sue is the name **paleontologists** gave to a famous *Tyrannosaurus rex* fossil. Sue was found in South Dakota. Her skeleton is the biggest *T. rex* ever found.

Dinoprints

Fossils include more than bones. Fossils can be created when an animal steps or lands in mud and leaves a mark called an *impression*. If that impression fills in with other material and stays filled in over time, it can become a fossil.

Paleontologists, scientists who study ancient life, work carefully to uncover fossils. They go slowly and use small tools to make sure nothing breaks.

How do we know about dinosaurs if they lived so long ago?

People know about dinosaurs because of fossils. Fossils form over many years, as the bones of a once-living thing turn to stone. Most of the time, scientists find fossil parts of an animal, but not its whole skeleton. They have to puzzle out how the pieces might have fit together.

STEAM & Me

Make a good first impression! Ask an adult to get you some clay and roll it out like a fat pancake. Gently press a shape into the clay—use a shell, a leaf, or even a toy. The shape left behind is an impression, just like a dinosaur footprint.

11

Dinosaurs lived here, there, and everywhere.

Dinosaur fossils have been discovered on every continent in the world. *Cryolophosaurus* was even found in chilly Antarctica! The United States, China, Mongolia, and Argentina have found the most dinosaur fossils so far.

Earth on the Move

One reason dinosaur fossils are found all over the world might be because the land on the globe used to look much different. Before 175 million years ago, all the land on Earth was smushed into one big continent called Pangaea. Slowly, over many years, that big landmass moved apart into the continents we know today.

Dino Diggers

Paleontologists use special tools to dig for dinosaur fossils. Some of those tools are chisels, geology hammers, and brushes. Sometimes they even use screwdrivers and tiny dental picks to scrape away small bits of rock.

Cryolophosaurus is one of the biggest dinosaur hunters, or predators. Scientists found *Cryolophosaurus*'s skull, which had a special crest, or bone, sticking out of it. Scientists aren't sure how *Cryolophosaurus* used its crest.

Teeny-Tiny Dinosaur
Microraptor: 22 inches long

Really BIG Dinosaur
Spinosaurus: 50 feet long

14

Argentinosaurus was a plant-eater. This Cretaceous-period dinosaur might have weighed more than 150,000 pounds.

Some dinosaurs were HUGE and some were small.

One of the biggest dinosaurs ever discovered is *Argentinosaurus*. It was more than 100 feet long. That's longer than two school buses parked end to end. *Compsognathus*, one of the smallest dinosaurs, was about 12 inches tall. That's shorter than a chicken.

STEAM

Measure up. *Compsognathus* was about 12 inches tall—the size of a ruler. From the bottom of your foot, use a ruler to measure how tall *Compsognathus* would be compared to you.

15

Made for Munching

Sauropods used their teeth like rakes to scoop up leaves and grasses to eat. *Triceratops'* teeth were flat and grooved, which was good for grinding the plants they ate.

Parasaurolophus was a hadrosaur, a type of duck-billed dinosaur. Its fossils were found in North America.

Some dinosaurs ate plants.

Many dinosaurs were **herbivores**, or plant-eaters. Big-horned dinos like *Triceratops* had long beaks for snipping plants. Hadrosaurs are nicknamed duck-billed dinosaurs because of their mouth shape, which helped them pull up water plants.

Eat Your Vegetables!

The biggest dinosaurs ate only plants! A sauropod is a type of dinosaur that walked on four legs and had a long neck and a long tail. *Apatosaurus*, *Brachiosaurus*, and *Camarasaurus* are all sauropods.

STEAM & Me

Say *ahhhhhh!* With a clean hand, feel the teeth in the back of your mouth. What do they feel like? The teeth in the back of your mouth are called molars. Molars are flat and grooved—like a *Triceratops*'s teeth!

Some dinosaurs ate meat.

Lots of dinosaurs were **carnivores**, or meat-eaters. Fierce dinos like *T. rex* and *Allosaurus* had huge, sharp teeth for biting into meat. They sometimes used their teeth for fighting.

Say *ahhhhhh* again! With a clean hand, feel the teeth in the front of your mouth. How do they feel different from the ones in the back of your mouth? Do you have any teeth like *T. rex*'s? The teeth in the front of your mouth are called incisors and canines. Incisors and canines are sharp, like a meat-eater's teeth.

STEAM & Me

T. rex would rip apart prey with its big teeth and then swallow without chewing. Gulp!

What Big Teeth You Have!

T. rex's teeth were about as big as bananas!

Eat Your Meat!

Theropods were a type of meat-eating dinosaur that walked on two legs. Some theropods like *T. rex* were big, and others were small, like *Compsognathus*.

Pachycephalosaurus was a bone-headed dinosaur. Really! Its skull was extra thick for headbutting.

Horns. Claws. Spikes. Armor. Dinosaurs had more than their teeth to protect them.

Meat-eaters and plant-eaters both had plenty of ways to defend themselves in a fight. Some of those features looked pretty wild. *Styracosaurus* had a large bony plate sticking up from its head, and that plate was covered with spikes.

Claws!

Deinonychus had huge claws on its feet. Its name means "terrible claw." *Ankylosaurus* had a bumpy shell, almost like a turtle's. This armor helped protect it from attack.

Spikes!

Stegosaurus had a spiked tail. Scientists think it might have swung the tail at predators.

Toot!
Some dinosaurs,
like the *Parasaurolophus*,
had large crests on their heads.
When scientists made a model
of *Parasaurolophus*'s crest,
it made a toot like
a tuba!

Talk like a dinosaur.

If dinosaurs made noise, they probably made grunting, groaning, or hissing sounds like today's reptiles do. Scientists also think dinosaurs might have communicated through body language like moving the head or arching the back.

Shake Those Tail Feathers

Some scientists think that dinosaurs had feathers. Dinosaurs may have used their feathers to communicate, just like birds do.

Dinosaurs on the move!

Some dinosaurs walked on two legs. *T. rex* had little arms but could run on its two very big and strong legs. Some dinosaurs, like *Triceratops*, walked on four legs and moved like a rhinoceros. Other dinosaurs could move on either two legs or four legs.

Iguanodon **walked on four legs but could also stand up on two legs.**

How fast can you go? Get down on your hands and knees. Crawl on the ground as fast as you can to the other side of the room. Now get up and walk the same distance. Are you faster on four legs or two?

STEAM&Me

Zoom!

Little *Compsognathus* might have been able to run at speeds up to 40 miles per hour, which is as fast as a motor scooter.

Where did all the dinosaurs go?

All the dinosaur fossils we've found are at least 65 million years old. So far, there are no dinosaur fossils from after that time. That means no more dinosaurs have lived on Earth since then. So after living on Earth for 145 million years, dinosaurs died out, or went **extinct**. What do you think happened to the dinosaurs?

Scientists have come up with a few ideas for why dinosaurs went extinct. One idea is that a huge asteroid—one wider than 100 football fields—crashed into Earth from space and destroyed the dinosaurs' environment.

Volcano!

A giant volcano may have erupted, sending ash clouds into the sky, blocking the sun so that the plants dinosaurs ate died out.

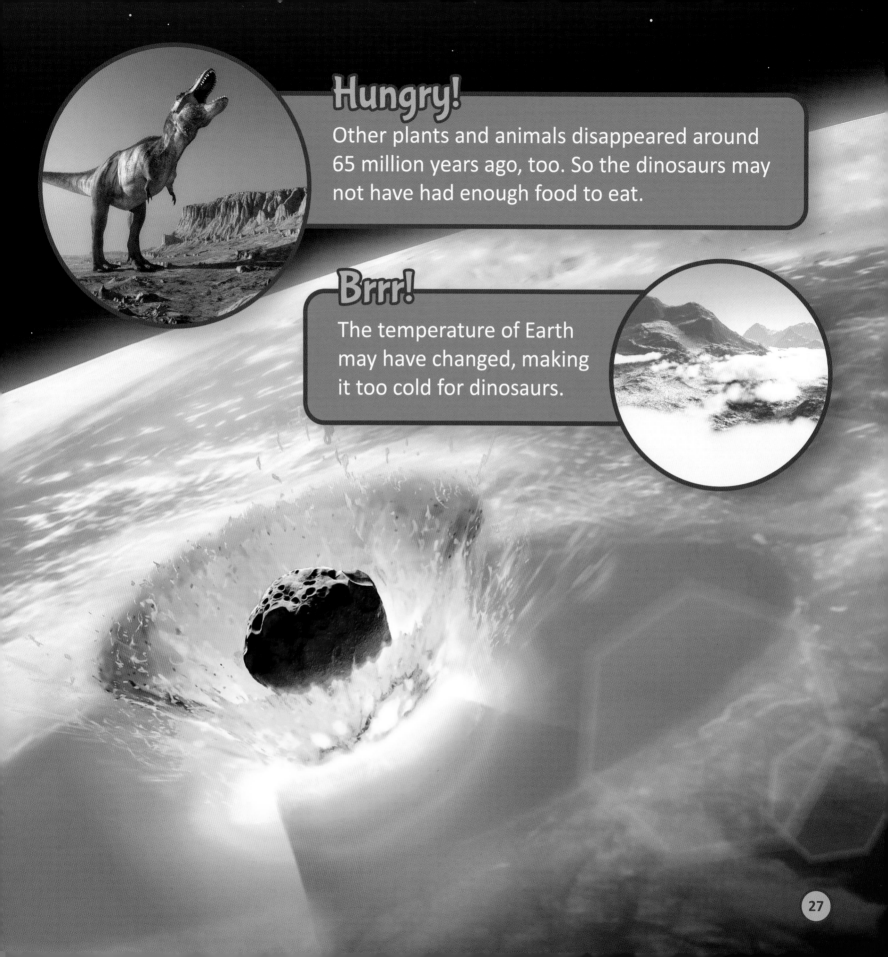

Hungry!

Other plants and animals disappeared around 65 million years ago, too. So the dinosaurs may not have had enough food to eat.

Brrr!

The temperature of Earth may have changed, making it too cold for dinosaurs.

Part bird, part dinosaur. Fossils of *Archaeopteryx* show that it was a little bit like a bird and a little bit like a dinosaur.

Dinosaur!
It had teeth like a dinosaur.

Bird!
It had wings like a bird.

Bird!
It had feathers like a bird.

Dinosaur!
It had a tail like a dinosaur.

Dinosaurs may have relatives around today.

Look outside in your yard. Do you see a bird? You may be looking at a dinosaur's relative! Some scientists think birds evolved from dinosaurs.

T. rex to Turkey
Scientists think birds evolved from theropods, which are two-legged, meat-eating dinosaurs, such as *T. rex* and *Velociraptor*.

So many dinosaurs!

Which dinosaurs here do you think are carnivores? Which are herbivores? Why? Combine the dino features you like best and create your own dinosaur. What will you name it? What does it eat? Where does it live?

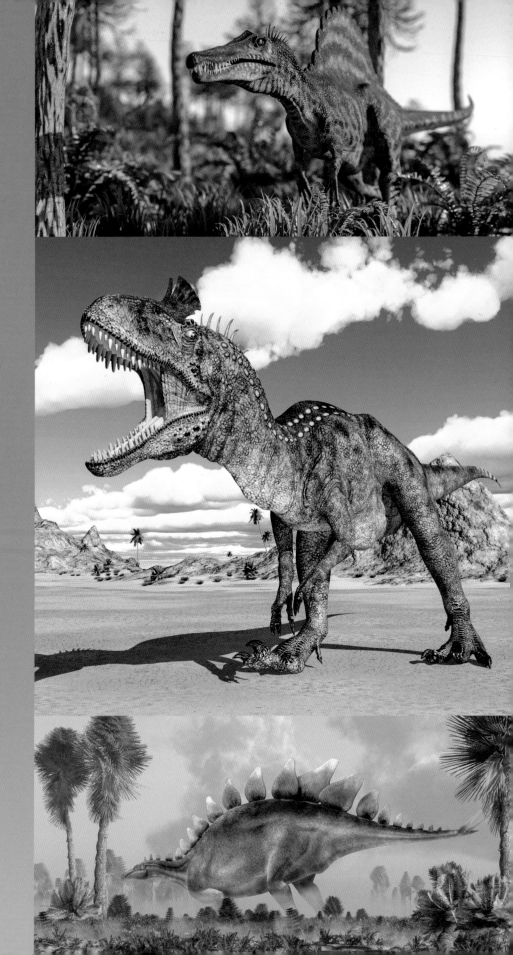

Glossary

Learn these key words and make them your own!

carnivore: a meat-eater. *T. rex was a* carnivore *and hunted other dinosaurs.*

era: a period of time. *Dinosaurs lived during the Mesozoic* era.

extinct: does not exist anymore. *Triceratops is* extinct, *so I can't see one at the zoo.*

fossil: evidence of an animal or a plant embedded in rock; can be bone, an outline, or a print. *Sue is the name of a T. rex pieced together from* fossils.

herbivore: a plant-eater. *A cow eats only plants, so it's an* herbivore.

paleontologist: a scientist who studies ancient life. *The* paleontologist *studies Iguanodon teeth.*

predator: an animal that hunts other animals. *Velociraptor was an* excellent predator *that hunted other dinosaurs.*

For Miles, a most dino-mite kiddo.

STEAM & Me and Starry Forest® are trademarks or registered trademarks of Starry Forest Books, Inc. • Text and Illustrations © 2020 and 2021 by Starry Forest Books, Inc. • This 2021 edition published by Starry Forest Books, Inc. • P.O. Box 1797, 217 East 70th Street, New York, NY 10021 • All rights reserved. No part of this publication may be reproduced, stored in a retrieval system, or transmitted in any form or by any means (including electronic, mechanical, photocopying, recording, or otherwise) without prior written permission from the publisher. • ISBN 978-1-946260-91-8 • Manufactured in China • Lot #: 2 4 6 8 10 9 7 5 3 1 • 03/21